I0115897

30 MAGICAL INSECTS WITH FASCINATING FACTS

© 2025 BY LINDA BLACKMOOR

ALL RIGHTS RESERVED. NO PORTION OF THIS BOOK MAY BE
REPRODUCED, STORED IN RETRIEVAL SYSTEM, OR TRANSMITTED IN ANY
FORM OR BY ANY MEANS - ELECTRONIC, MECHANICAL, PHOTOCOPY,
RECORDING, SCANNING, OR OTHER - EXCEPT FOR BRIEF QUESTIONS IN
CRITICAL REVIEWS OR ARTICLES, WITHOUT THE PRIOR WRITTEN
PERMISSION OF THE PUBLISHER OR AUTHOR.

ISBN: 978-1-966417-29-3 (PRINT)

PUBLISHED BY QUILL PRESS. LINDA BLACKMOOR'S TITLES MAY BE
PURCHASED IN BULK FOR EDUCATIONAL, BUSINESS, FUNDRAISING, OR
SALES PROMOTIONAL USE. FOR INFORMATION, PLEASE EMAIL
HELLO@LINDABLACKMOOR.COM

FIRST PRINT EDITION: 2025

LINDA BLACKMOOR
WWW.LINDABLACKMOOR.COM

LUNA MOTH

Scientific Name: Actias Luna

Location: Eastern North America

Size:
3 to 4.5
inch wingspan

Color:
Pale green with
eye spots

Fact:
Luna Moths have no
mouths—they never
eat as adults.

Lifespan:
7-10 days as
an adult

Food Source:
Tree leaves
as
caterpillar;
none as adult

#1

BUMBLE BEE

Scientific Name: Bombus spp.

Location: North America, Europe, Asia, South America

Size:
0.4 to 1
inch (1 to 2.5 cm)

Color:
Black and yellow with
fuzzy bodies

Fact:
Bumble bees can fly in
chilly weather when
other bees can't.

Lifespan:
Workers live
a few weeks;
queens up to
1 year

Food Source:
Nectar and
pollen from
flowers

#2

PRAYING MANTIS

Scientific Name: Mantodea, Tenodera Sinensis

Location: Worldwide Except Antarctica

Size:
1 to
6 inches,
depending on
species

Color:
Green, brown,
pink, white,
yellow, and even
shades of blue

Fact:
Some mantises mimic
flowers so well they
lure in prey with
beauty.

Lifespan:
6 months to 1
year

Food Source:
Insects,
spiders, small
frogs, lizards,
and even
hummingbirds

3

LADY BUG

Scientific Name: Coccinellidae

Location: Worldwide Except Antarctica

Size:
0.3 to
0.4 inches
(7 to 10 mm)

Color:
Red or orange with
black spots, yellow,
pink, black, gray, and
shiny metallic blue

Fact:
A single ladybug
can eat up to 5,000
aphids in its lifetime.

Lifespan:
About 1 year

Food Source:
Aphids,
mites, scale
insects, and
insect eggs;
some species
eat plants or
mildew

4

MONARCH

Scientific Name: Danaus Plexippus

Location: American Continents, Australia, Europe, Asia

Size:
3.5 to 4
inches wingspan

Color;
Bright
orange with black
veins and white spots

Fact:
Monarchs migrate up
to 3,000 miles.

Lifespan:
7-2 to 6 weeks;
up to 8 months for
migrating
generation

Food Source:
Milkweed (as
caterpillar);
nectar from flowers
(as adult))

#5

A N T

Scientific Name: Formicidae

Location: Worldwide Except Antarctica

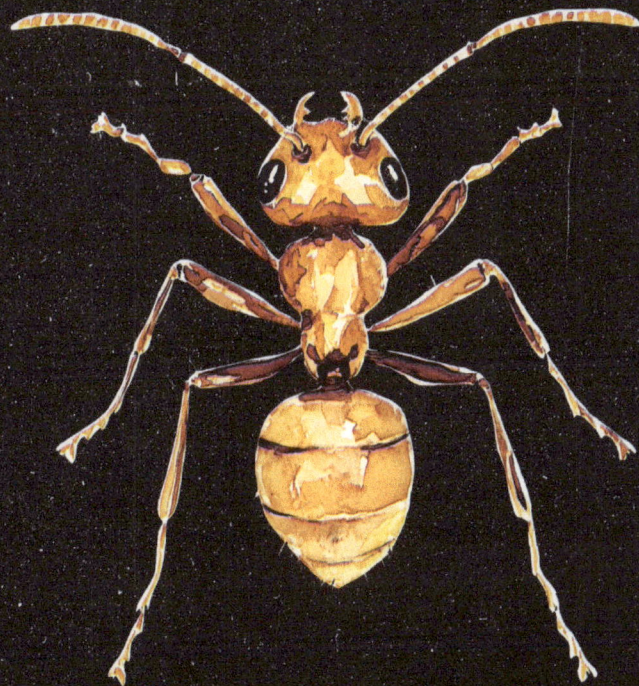

Size:
0.03 to
2 inches

Color:
Black, red, brown,
yellow, green or
metallic

Fact:
Ants can carry 10 to
50 times their own
body weight.

Lifespan:
Workers live
weeks to months;
queens can live up
to 30 years

Food Source:
Sugars, proteins,
fungi, seeds,
insects, and even
other ants

6

GRASSHOPPER

Scientific Name: Caelifera

Location: Worldwide Except Antarctica

Size:
0.5 to
4 inches

Color:
Green, brown,
yellow, pink;
striped or
patterned

Fact:
Grasshoppers
hear with ears
on their bellies.

Lifespan:
Several weeks to
a few months

Food Source:
Leaves, grasses,
crops, and other
plant material

7

CICADA

Scientific Name: Cicadidae

Location: Worldwide Except Antarctica

Size:
1 to 2.5
inches

Color:
Brown, black, green;
clear or veined wings

Fact:
Males sing using
vibrating membranes
in their abdomen.

Lifespan:
2 to 5 years (some
live 13 or 17 years
underground as
nymphs)

Food Source:
Leaves, grasses,
crops, and other
plant material

#8

DRAGONFLY

Scientific Name: Anisoptera

Location: Worldwide Except Antarctica

Size:
1 to 5 inches

Color:
Metallic blues,
greens, reds, yellows,
and black—often
iridescent

Fact:
Dragonflies can fly in
all directions—even
backward.

Lifespan:
Up to 6 months as
adult; several years
as aquatic nymph

Food Source:
Mosquitoes, flies,
moths, and other
small flying
insects

9

SCARAB BEETLE

Scientific Name: Scarabaeidae

Location: Worldwide Except Antarctica

Size:
0.2 to
2.5 inches

Color:
Black, brown, metallic
green, gold, or
iridescent rainbow

Fact:
Some species roll
dung into
perfect balls.

Lifespan:
Several months
to 1 year

Food Source:
Dung, decaying
plants, fungi,
fruit, or roots
(varies by species)

#10

WOOLLY BEAR

Scientific Name: Pyrrharctia Isabella

Location: North and Central America

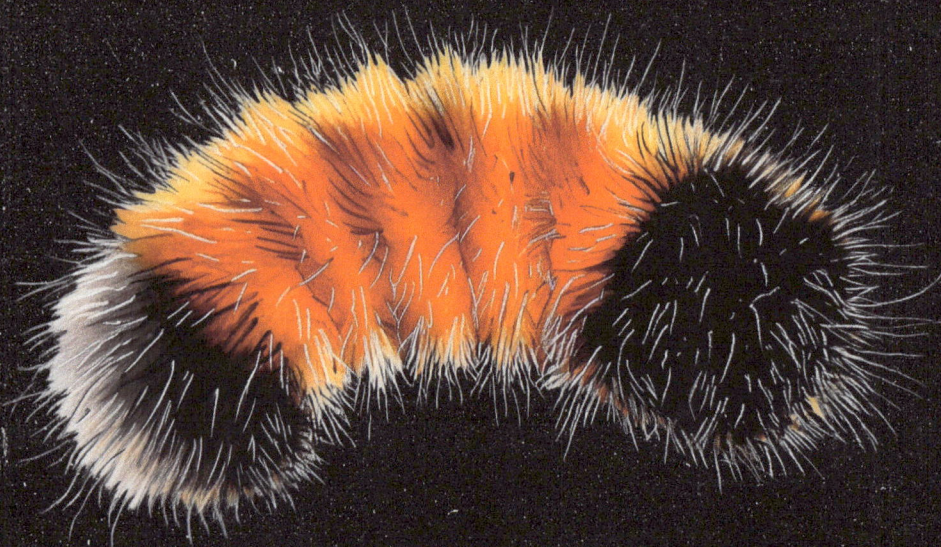

Size:
About 2
inches

Color:
Black at both ends
with an orange band in
the middle

Fact:
Woolly Bears can
survive freezing
solid in winter.

Lifespan:
Several weeks as
a caterpillar; 1
year total life
cycle

Food Source:
Leaves of low-
growing plants like
dandelions, clover,
and plantain

#11

HONEY BEE

Scientific Name: Apis Mellifera

Location: Worldwide Except Extreme Cold Regions

Size:
About 0.5
inches

Color:
Golden brown
with black stripes
and fine hairs

Fact:
Honey bees use a
"waggle dance" to
show others where
to find food.

Lifespan:
5 to 6 weeks
(worker), up to 5
years (queen)

Food Source:
Nectar and pollen
from flowers

#12

BLACK WIDOW

Scientific Name: Latrodectus Mactans

Location: North and South America

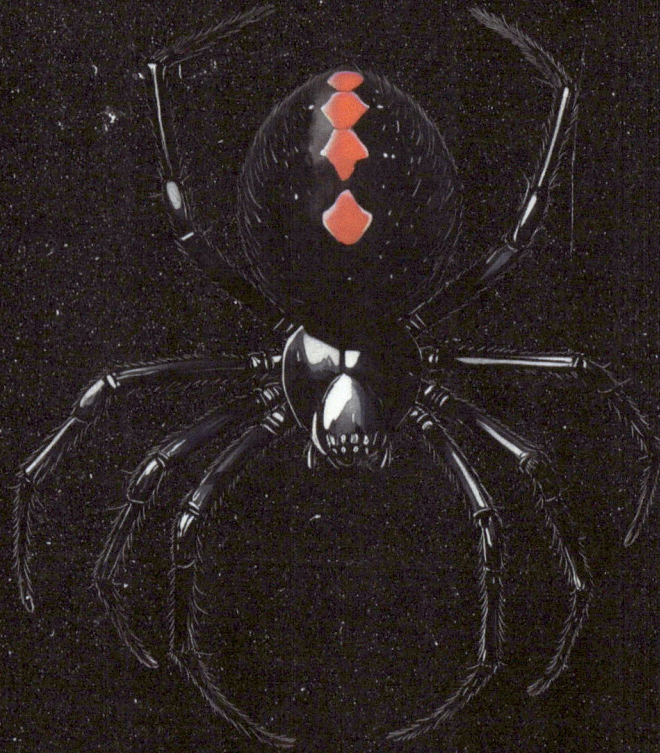

Size:
1.5 inches
with legs
extended

Color:
Shiny black
with a red
hourglass

Fact:
Only females
have the
hourglass,
found on the
underside.

Lifespan:
1 to 3 years

Food Source:
Insects and
other small
arthropods
caught in
their webs

#13

COCKROACH

Scientific Name: Blattodea
Location: Worldwide

Size:
0.5 to
3 inches

Color:
Brown, black,
reddish; Shiny or
patterned shells

Fact:
Cockroaches can
live for a week
without their
heads.

Lifespan:
Several
months to over
a year

Food Source:
Almost
anything—
crumbs, paper,
glue, garbage,
and more

#14

JAPANESE BEATLE

Scientific Name: Popillia Japonica

Location: Japan; Invasive in North America and Europe

Size:
About 0.5 inches

Color:
Metallic green
body with coppery
wing covers

Fact:
Japanese beetles
release scents to
call in swarms for
feeding.

Lifespan:
About 30 to 45
days

Food Source:
Leaves,
flowers, and
fruits of over
300 plant
species

#15

SILVERFISH

Scientific Name: Lepisma Saccharina

Location: Worldwide Except Antarctica

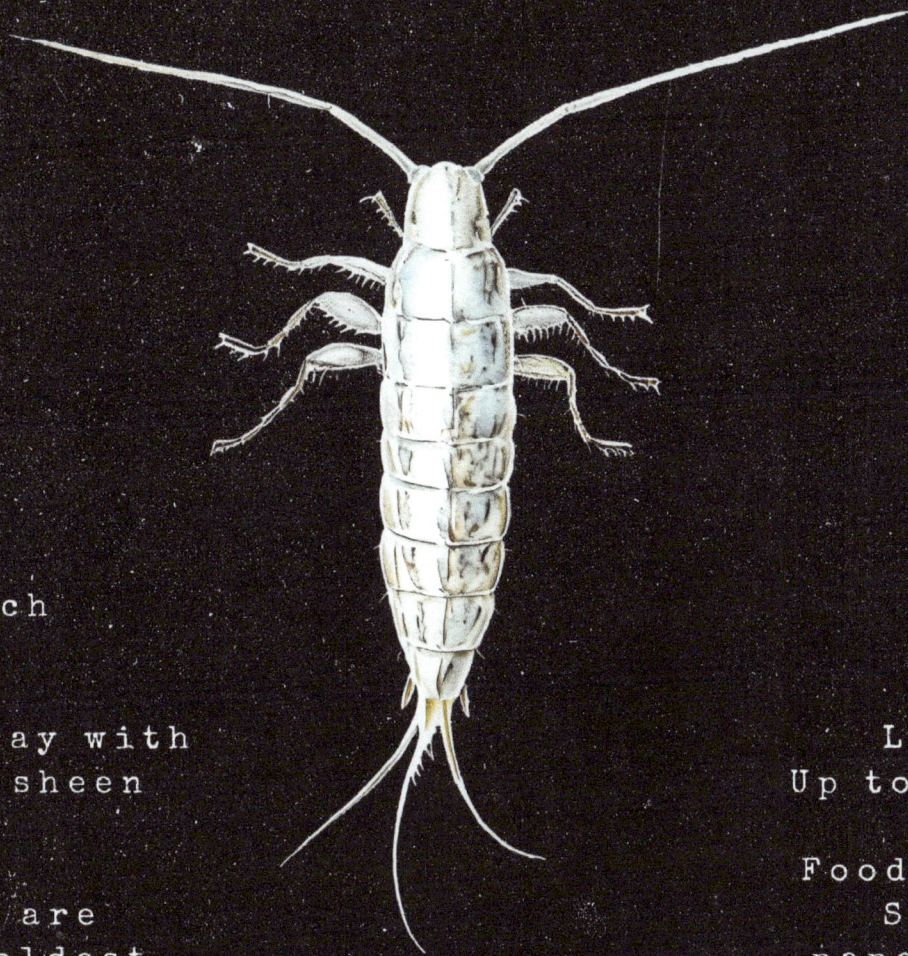

Size:
0.5 to 1 inch

Color:
Silvery-gray with
a metallic sheen

Fact:
Silverfish are
among the oldest
insects on Earth—
over 400 million
years old.

Lifespan:
Up to 8 years

Food Source:
Starches,
paper, glue,
fabrics, and
even dead
insects

#16

ROSY MAPLE MOTH

Scientific Name: Dryocampa Rubicunda
Location: Eastern North America

Size:
1 to 2 inches
wingspan

Color:
Bright pink
and yellow
with fuzzy
bodies

Fact:
Despite their
bright colors,
they are
masters of
camouflage
against spring
flowers.

Lifespan:
2 to 9 months
as pupa;
adults live
about 1 week

Food Source:
Maple and oak
leaves as
caterpillars;
adults do not
eat

#17

CRICKET

Scientific Name: Gryllidae

Location: Worldwide Except Antarctica

Size:
0.5 to 2
inches

Color:
Brown, black,
green, or
yellow

Fact:
Male crickets
chirp by
rubbing their
wings together
—not their
legs.

Lifespan:
6 weeks to 3
months

Food Source:
Plants,
fungi,
decaying
matter, and
small insects

#18

TICK

Scientific Name: Ixodida

Location: Worldwide Except Antarctica

Size:
0.1 to
0.3 inches

Color:
Brown,
reddish-brown,
or black

Fact:
Ticks can go
months—even
years—without
a meal.

Lifespan:
2 months to 3
years

Food Source:
Blood from
mammals,
birds,
reptiles, and
amphibians

#19

RED VELVET ANT

Scientific Name: Dasymutilla Occidentalis

Location: Southern and Eastern United States

Size:
0.5 to
1 inch

Color:
Bright red-
orange and black
with a fuzzy body

Fact:
It's not an ant
at all—it's a
wingless wasp
with one of the
most painful
stings in North
America.

Lifespan:
1 to 2 years

Food Source:
Nectar
(adults);
larvae feed
on the pupae
of ground-
nesting bees
and wasps

#20

STINK BUG

Scientific Name: Pentatomidae

Location: Worldwide Except Antarctica

Size:
0.5 to
0.75 inches

Color:
Mottled brown,
green, or gray

Fact:
Stink bugs
release a smelly
chemical when
scared.

Lifespan:
6 to 8 months

Food Source:
Plant sap
from fruits,
vegetables,
and leaves

#21

MOSQUITO

Scientific Name: Culicidae

Location: Worldwide Except Antarctica

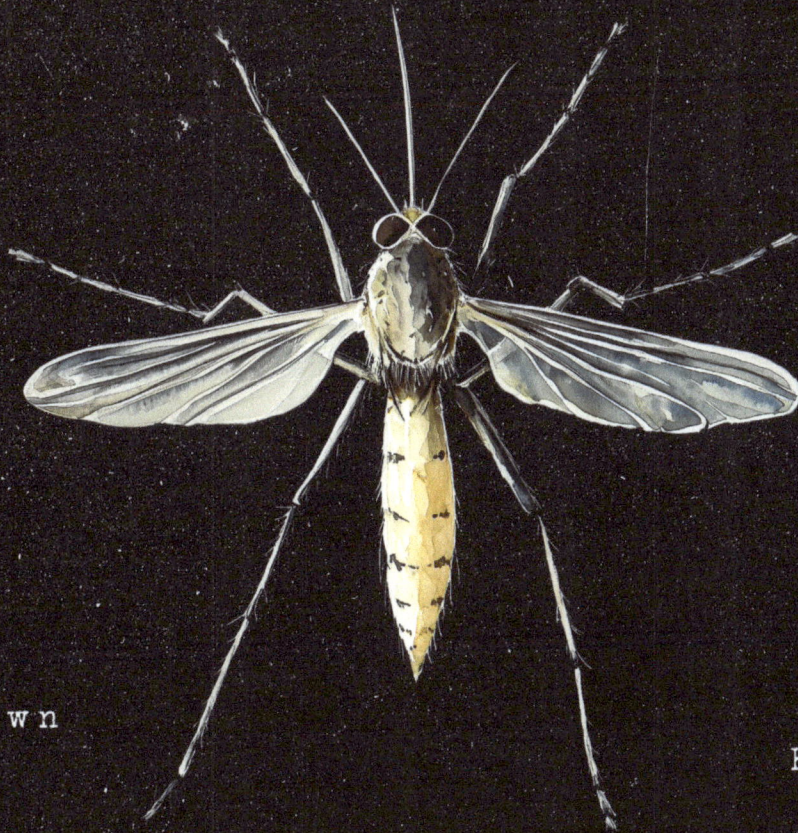

Size:
0.1 to 0.4
inches

Color:
Gray or brown

Fact:
Only female
mosquitoes bite
—and they can
smell you from
over 100 feet
away.

Lifespan:
2 to 4 weeks

Food Source:
Nectar
(males);
female
mosquitoes
drink blood
to develop
eggs

#22

YELLOWJACKET

Scientific Name: Vespula and Dolichovespula

Location: North America, Europe, and parts of Asia

Size:
0.5 to 0.75
inches

Color:
Bright yellow
and black

Fact:
Yellowjackets
can sting
multiple times—
and they get more
aggressive in
late summer.

Lifespan:
Workers live
a few weeks;
queens can
live up to a
year

Food Source:
Sugary foods,
nectar, and
other insects

2 3

HOUSE FLY

Scientific Name: Musca Domestica

Location: Worldwide Except Antarctica

Size:
About 0.25 inches

Color:
Gray body with red eyes and transparent wings

Fact:
House flies taste with their feet— and vomit on food to dissolve it before eating.

Lifespan:
15 to 30 days

Food Source:
Rotting food, garbage, feces, and sugary substances

#24

SWALLOWTAIL

Scientific Name: Papilionidae

Location: Worldwide Except Antarctica

Size:
3 to 6 inches
wingspan

Color:
Often yellow,
black, blue, or
green; multi-
colored

Fact:
Swallowtail
caterpillars can
puff out orange
"horns" to scare
predators.

Lifespan:
About 1 month

Food Source:
Nectar from
flowers;
caterpillars
feed on
plants like
parsley,
dill, and
citrus

#25

JUMPING SPIDER

Scientific Name: Salticidae

Location: Worldwide Except Antarctica

Size:
0.1 to 0.8
inches

Color:
Black, brown,
orange, white,
metallic blue or
green

Fact:
Jumping spiders
can leap up to 50
times their body
length.

Lifespan:
About 1 year

Food Source:
Small
insects
and other
spiders

#26

PILL BUG

Scientific Name: Armadillidium Vulgare

Location: Worldwide Except Antarctica

Size:
0.3 to 0.7
inches

Color:
Dark gray or
brown

Fact:
Pill bugs aren't
insects—they're
crustaceans, more
closely related to
shrimp than bugs.

Lifespan:
About 2 to 5
years

Food Source:
Decaying
plant matter,
dead leaves,
wood, and
fungi

#27

KATYDID

Scientific Name: Tettigoniidae

Location: Worldwide Except Antarctica

Size:
1 to 4
inches

Color:
Usually green,
sometimes
brown or pink

Fact:
Katydids can hear
with ears on their
front legs—tiny
slits that pick up
high-pitched sounds.

Lifespan:
About 1 year

Food Source:
Leaves,
flowers,
bark, and
occasionally
other insects

#28

STICK INSECT

Scientific Name: Phasmatodea

Location: Worldwide Except Antarctica

Size:
1 to 13
inches

Color:
Brown, green,
gray; some
species have
white, pink,
red, or blue

Fact:
Some stick insects
can regrow lost legs.

Lifespan:
1 to 2 years

Food Source:
Leaves from
trees and
shrubs

#29

LEAF INSECT

Scientific Name: Phylliidae

Location: Southeast Asia, Australia, Nearby Islands

Size:
2 to 4
inches

Color:
Green or
yellow-green

Fact:
Females are
usually bigger
and leafier—males
are smaller and
can fly.

Lifespan:
5 to 8 months

Food Source:
Leaves from
trees and
shrubs like
guava, mango,
and bramble

#30

SHOP MORE TITLES:

WWW.LINDABLACKMOOR.COM

which WITCH am i
BEDTIME STORY FOR LITTLE WITCHES
LINDA BLACKMOOR

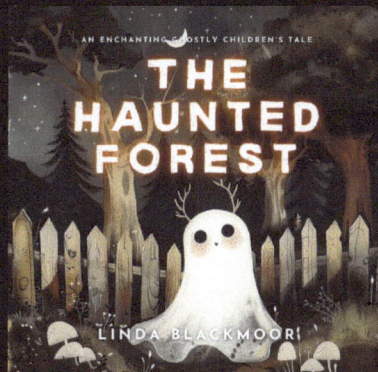

AN ENCHANTING GHOSTLY CHILDREN'S TALE
THE HAUNTED FOREST
LINDA BLACKMOOR

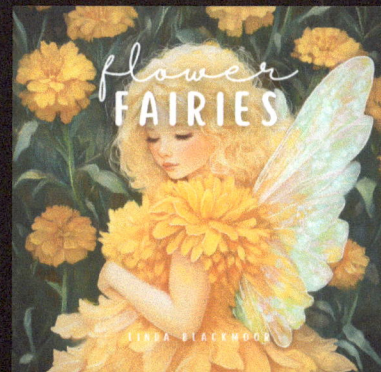

flower FAIRIES
LINDA BLACKMOOR

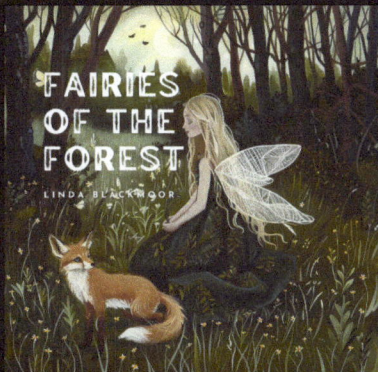

FAIRIES OF THE FOREST
LINDA BLACKMOOR

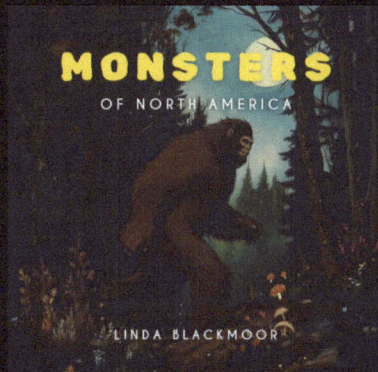

MONSTERS OF NORTH AMERICA
LINDA BLACKMOOR

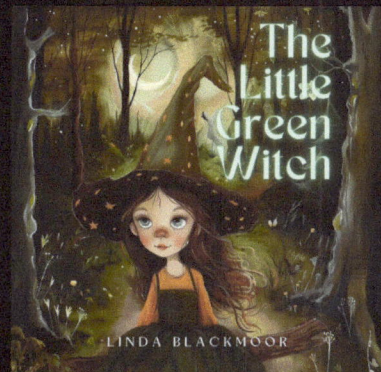

The Little Green Witch
LINDA BLACKMOOR

www.ingramcontent.com/pod-product-compliance
Lightning Source LLC
Chambersburg PA
CBHW060834270326
41933CB00002B/87

* 9 7 8 1 9 6 6 4 1 7 2 9 3 *